THE MULTIMODAL WORLD

AUDREY E. RANDLES

2020

THE MULTIMODAL WORLD

Copyright © 2020 AUDREY ELIZABETH RANDLES
All rights reserved.
ISBN: 9798643380689

Cover Image: 'North America Nebula in Different Lights'
Image credit: NASA/JPL-Caltech

THE MULTIMODAL WORLD

CONTENTS

1	Acknoledgement	Pg 4
2	Introduction	Pg 6
3	The Matrix	Pg 9
4	Energy draws space and time	Pg 17
5	Galaxies and star clusters	Pg 20
6	Binary and multi-systems	Pg 26
7	STIE-blocks	Pg 36
8	The multimodal Universe	Pg 42
9	Four principles of the objects existence	Pg 49
10	Prospects	Pg 51
11	About the author	Pg 53
12	The list of the images	Pg 55

THE MULTIMODAL WORLD

ACKNOWLEDGEMENT

We would like to express our gratitude to the National Aeronautics and Space Administration (NASA), NASA's Jet Propulsion Laboratory (JPL), California Institute of Technology (Caltech), the Space Telescope Science Institute (STScI), the University of California at Los Angeles (UCLA), Vassar College (Vassar), Digitized Sky Survey (DSS), and the Max Planck Institute for Astronomy (MPIA) for the Space images and exciting descriptions, published by NASA's Jet Propulsion Laboratory (JPL), and used in this book to illustrate the spectacular beauty of our dynamic world, its complex structures, and success of the space explorations addressing fundamental questions of the Earth physics and human psychology concerning our place in the Universe. We wish to thank the Oxford Dictionary for the precise formulation of the tendencies existing in the modern natural sciences.

The views and opinions, expressed in this book, do not necessarily state or reflect those of the Oxford Dictionary, NASA's Jet Propulsion Laboratory (JPL), California Institute of Technology (Caltech), the Space Telescope Science Institute (STScI), the University of California at Los Angeles (UCLA), Vassar College (Vassar), Digitized Sky Survey (DSS), the Max Planck Institute for Astronomy (MPIA), and the National Aeronautics and Space Administration (NASA).

Image credit: NASA/JPL-Caltech

INTRODUCTION

I would like to introduce the Theory of Matrix by re-phrasing the words of Hermann Minkowski. The views of Space and Time, which I wish to present, have sprung from the soil of experimental psychology. 'Herein lies their strength. Their tendency is radical.'

The theory was born in the early 1990s. It was developed along with the Coresynthesis Psychological Model as a new theory of Self and Reality. The propositions of the Theory of Matrix would not start from the concept of a thinking-being in general, but from the reality we operate in.

In the Theory of Matrix, we introduce a new understanding of the world as the multimodal world. The multimodal world is the self-regulating multimodal Space-Time and Into-Energy system.

Time, space, mass, and different forms of energy do not exist independent of objects and systems. They are the properties of the existing objects and systems of the objects. The volumes, masses, forces, energy, information, and other characteristics of the existing objects and systems are reflected in the properties of the Universe.

Our thermodynamic Universe is one of the finite modalities of the Great Universe. It is limited in time, space, mass, energy, and associated information. The Universe, as we know it, is expanding in space. It is a young non-radiating Universe, and it is likely to develop a toroidal form in the future.

The complement to our dynamic world is the radiating modality of the Great Universe. It is a source of space, time, matter, and energy, sustaining our dynamic world.

Other modalities of the Universe include Multiverse, other Universes, and stages of the Universe development. Every modality of the Universe has the start and the end in time and space.

The properties of the Universe modalities are integrated and reflected in the characteristics of the Great Universe existing in the state of equilibrium.

We present the main principles, associated with the multimodal Space-Time structure of the existing objects and systems, including Black Holes, visually 'empty' spaces, stars and galaxies, subatomic particles and holes, our thermodynamic Universe and other Universe modalities.

Energies of background radiation and associated information, supporting the multimodal objects and systems, make them the 'space-time-info-energy' building blocks, or STIE-blocks of the Universe. Accordingly, background radiation, including those currently known as cosmic background radiation, and light set the background structure of our dynamic world.

Image credit: NASA/JPL-Caltech/MPIA

THE MATRIX

Human perception of time as a time-point, the current 'now', is the main boundary we operate in our dynamic world. This boundary is a result of the evolutionary stage, limiting the human perception of the multimodal world. Consciously we live 'now', at a '0' time-point dividing the past from the future.

We split the world, separate it, extrapolate the qualities, spaces, energies, forms, changes by inferring unknown values and trends from the whole of the Universe. We build evidence, skills, and judgements, in-forming our conscious and unconscious processes based on our senses, using our bodies, fixing and coding the forms and experiences within our memory.

The world is real. Matter, energy, forms, time, and space really exist, but we can deal with only a very limited modality of the Great Universe - our dynamic world as we know and sense it. The complement, or, say, the other multiple modalities of the Universe are currently not accessible to humans.

A '0' Space-Time point is the point at which our perception and consciousness begin to diverge from the reality of the multimodal world. A '0' Space-Time point, or 'a space-time-null-point', is a point 'here' and 'now'. The '0' Space-Time point may be represented as a coexistence of the '0' time-point - 'now' and '0' space-point - 'here', which are complementary, building the time system and the space system in the opposite direction.

The point 'here' and 'now' as a '0' point was mentioned in the Theory of Special Relativity in association with the mathematical model of the Light Cone. 'Let us call any world-point 0 as a space-time-null-point' [Minkowski Hermann, 'Space and Time' (1920)]. 'A world-point is a 'here-now' [Weyl Hermann, 'The Discussion concerning the Theory of relativity at the Meeting of Natural Scientists' (1920)].

The '0' Space-Time point is the centre of the Matrix. In the Theory of Matrix, the word 'Matrix' means the multimodal Space-Time and Info-Energy structure of an object or a system. The Matrix reflects the human perception of time as a point 'now', associated with the period of time now occurring.

The description of the Matrix includes the typical characteristics introduced mathematically by Lorenz, Minkowski, and Einstein for the famous Light Cone, mainly associated with the Matrix symmetry, and some additional aspects found in our psychological investigations.

The '0' Space-Time point is the point of the Matrix symmetry. The Matrix Space-Time, Mass-Energy, and Info-Energy are symmetrical via the '0' Space-Time point if the Matrix is balanced. The '0' Space-Time point is a point of Space-Time transformations, keeping space and time in a dynamic balance under the Space-Time, Mass-Energy, and Info-Energy Conservation Laws and Matrix Laws of Symmetry.

According to the Theory of Special Relativity, the space axis of the Light Cone builds a perpendicular to the time axis. In compliance with our investigations of the Matrix and according to the Theory of Matrix, the Light Cone is an example of the Matrix for a flash of light and the Space-Time axis (other terms: Matrix axis, time axis, space axis) is the only axis of the Matrix. The Matrix Space-Time, Mass-Energy, and Info-Energy are symmetrical via the Space-Time axis if the Matrix is balanced.

The divergence of the sensuous actuality of our dynamic world from the reality of the Great Universe continues since the Space-Time Arrow, directed by the Space-Time axis, demonstrates a coexistence of four tendencies - the tendency of time and the contra-directed tendency of space enabling our perception of movement, along with the tendencies of the past and the future.

Space and time blend closely. Time influences the direction of the space 'flow'. A clock in a moving frame is seen to be 'dilated' according to the Lorentz transformation. Time is always 'dilated', and the length of a system shortens in our dynamic world relative to the stationary position of the system and relative to the stationary observer.

The Matrix reflects the current human understanding of two separate, directly imperceptible periods related to the past and the future. The time component of the Space-Time Arrow forms the Arrows of the Progressive and Regressive Time. Two Arrows of Time and two contra-directed Arrows of Space reflect the time and space components of the Space-Time Arrow. They depict the Matrix tendency of time and the contra-directed tendency of space reflected in our perception of speed.

Space-Time is represented in the Matrix as the Space of the Current Time associated with the period of time now occurring, Space of the Progressive Time related to the object's future, and Space of the Regressive Time associated with the object's past. Two Arrows of Time, associated with the Spaces of the Progressive and Regressive Time, are directed by the Matrix axis. The Arrows of Time indicate the direction of the time and time-associated energy 'flow' in the Matrix Spaces of Time. They are balanced and symmetrical if the Matrix is balanced in its Space-Time, Info-Energy, and Mass-Energy.

Two types of Matrixes have been identified - the Time-Rising Matrix (Figure 1) and the Space-Rising Matrix (Figure 2).

The direction of time and energy flow within radiating objects and systems is different from the time and energy direction in non-radiating objects and systems.

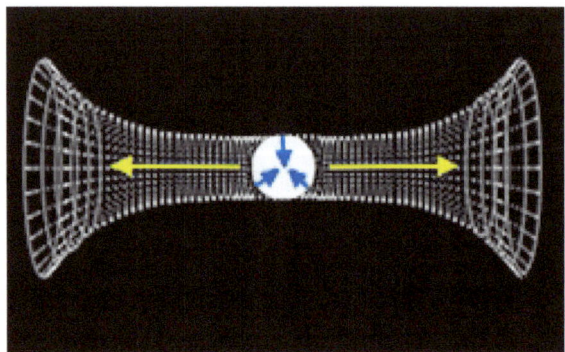

Figure 1: Time Rising Matrix

The Time-Rising Matrix (TRM) is a property of the radiating objects and systems, such as stars, planets, radiating star clusters and galaxies tangled by gravity. The TRM has a form of the Riemannian Manifold with the negative curvature (Figure 1).

The Space-Rising Matrix (SRM) is a property of the non-radiating objects and systems, such as the systems holding gravitating Black Holes and the Universe developing antigravity prompted by their periphery. The SRM has a form of the Riemannian Manifold with the positive curvature (Figure 2).

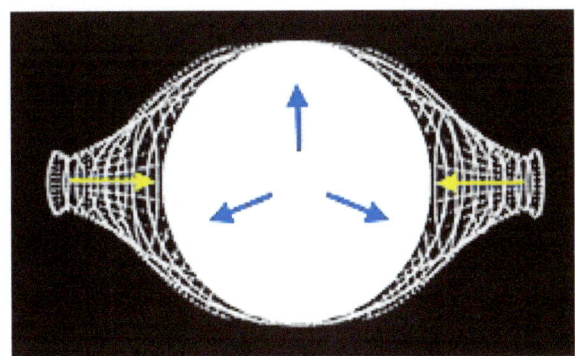

Figure 2: Space Rising Matrix

Arrows of Time are drawn as the yellow arrows, while the direction of the Space of the Current Time is represented by the blue arrows on the reproductions of the Matrixes (Figures 1 and 2). The Matrix Arrows of Space are situated in accordance with the human perception of the objects and systems as exhibiting three spatial dimensions (x, y, and z, or a combination of three directions, which can be chosen from the terms: length, width, height, depth, and breadth). It applies to all types of Matrixes.

Accordingly, the direction of the Arrows of Time in the Time Rising Matrixes (TRMs) of the radiating objects and systems is different from the direction of the

Arrows of Time in the Space Rising Matrixes (SRMs) of the non-radiating objects and systems. They display some specific characteristics associated with the type of the Matrix.

The direction of the Arrows of Time provides us with an opportunity to detect possible changes of the Space-Time direction and reverse of the energy flow within the objects and systems of the objects, for example, the Sun, our planet, Black Holes, and the Universe. Please see my book 'Energy of Existence' for details.

There is no priority of the Progressive or Regressive Time in the Matrix. The Matrix Spaces of the Progressive and Regressive Time are symmetrical via the Matrix centre and the Space-Time axis if the Matrix is balanced. Every Matrix displays a tendency to obtain and retain a balance and symmetry.

The Spaces of the Progressive and Regressive Time, formed by the 2-dimensional Matrix grid, look like a hologram. They are immobile, unchanging, inaccessible, appear empty and connected with the undifferentiated Continuum. The one-dimensional Space-Time Continuum is the one quality unified Space-Time of ultimate dynamics of the infinite duration.

Strictly speaking, the reproduction of a Matrix grid on the paper or the screen (Figures 1 and 2) is not exact though the presented forms of the Matrixes are correct. It is related to the difficulties to reproduce the 2-dimensional structures on the paper. It is a theoretical mistake that one can currently reproduce a 2-dimensional structure on the paper and PC screen. Our perception of the world influences any drawing and use of programming languages. A drawing on paper or PC screens, one layer of cells, crystals or atoms don't reproduce a 2-dimensional structure but project the 3-dimensional structure on the surface.

The Space of the Current Time, filled with the actual Info-Energy, equals the volume of the object enclosed at the Matrix centre. The term 'Info-Energy' means information and energy, which are complementary and balanced in the Matrix. Energy exists in a form determined by information, and information has a structure set by energy. Accordingly, information takes different forms and exists in the latent, or potential, and actual forms along with energy. The actual Info-Energy is represented in mass, kinetic energy, and other energy representations acting in the volume of the object in the period of time now occurring.

Image credit: NASA/JPL-Caltech

ENERGY DRAWS SPACE AND TIME

Energy draws space and time, objects and systems, forming our mental image of the world. Every existing object and every system of the objects with the characteristics of volume, mass, energy, and time of existence, represent an individual arrangement for our senses. Every existing object and every system have the original Matrix representing the total Info-Energy of this object or the system in space and time, including its past, present, and future.

According to Albert Einstein, 'the ponderable masses will be the determining factor in producing the field, or, according to the fundamental result of the special theory of relativity, the energy density...' [Albert Einstein, A Brief Outline of the Development of the Theory of Relativity (1921)].

At the same time, Albert Einstein had specified the connection between mass and energy of the system. 'If an amount of energy E be given to a body, the inertial mass of the body increases by an amount E/c^2, where c is the velocity of light in vacuo. On the other hand, a body of mass m is to be regarded as a store of energy of magnitude mc^2.' [Einstein Albert, A Brief Outline of the Development of the Theory of Relativity (1921)].

Space-Time is built by and filled with energy. Masses are the determining factor in producing the energy density, information density, and multimodal Space-Time and energy structures of their Matrixes. The objects and systems of limited mass and energy are limited in space and time. Accordingly, the Matrixes for the objects and systems of finite mass and energy are limited in space and time. Then again, an object of a finite volume is limited in mass, energy, and time of existence.

According to the Theory of Matrix, the Matrix is not the field but the structure. It is formed by the regular repeated, typically rectangular, grid-like two-dimensional Info-Energy arrangement - the Matrix grid.

The Matrix grid is an object's 2-dimensional framework holding a complete set

of data related to the enclosed object, in a resemblance to a chromosome of a living cell. It carries information in the form of Info-Energy blocks. Please see my book 'Energy of Existence' for details.

The Matrix grid, being the external framework of the object in the multimodal Space-Time of the Matrix, is the internal framework and Space-Time and Info-Energy skeleton of this object in our dynamic world.

The Matrix grid, arranged by the 2-dimensional representations of background radiation, is supported by energies of background radiation. Any object, which is smaller than the wavelength supporting the Matrix grid, does not exist in Space-Time.

The object is a source of its Matrix and, nevertheless, the Matrix is a self-regulating system. Matrixes are generated as an outcome of the binary operation of two Matrixes or by any of the following: reflection, self-duplication, binary fission, self-multiplication, and incorporation while maintaining anisotropy.

GALAXIES AND STAR CLUSTERS

Self-regulating systems are found in nature. We can arrange the star clusters and galaxies into three groups and analyse their behaviour. In our expanding Universe, we detect the systems of millions or billions of stars, gas and dust, which are held together by gravity, such as many Globular clusters filled with old stars.

Globular clusters, such as Omega Centauri - the largest globular cluster in the Milky Way, have spherical shapes (Image 12). They are bound by gravity and have higher densities toward their centres.

According to the Theory of Matrix, these star clusters and galaxies are 'old' in their energy scale. The type 1 class includes the radiating star clusters and galaxies expending in time and associated with their Time-Rising Matrixes. These systems have gone through the stage of being toroidal systems in the past. Currently, they perform the degeneration of toroid and develop time-associated latent Info-Energy structures, along with the consolidation of spaces and generation of mass reflected in gravitational processes prompted by their periphery. The balancing anti-gravitational processes and Black Holes, reflecting space and mass degeneration and reduction, might be detected in their central regions.

The type 2 class includes spacious and Space-Rising non-radiating systems, expanding into flattened and diluted space. They are associated with their Space-

Rising Matrixes. These systems are likely to develop a toroidal form in the 2-dimensional Potential Space in the future. Gravitational processes, reflecting space and mass generation, could be detected at the centres of these systems. The balancing anti-gravitational processes, reflecting space and mass degeneration and reduction, might be identified in their peripheral areas.

The type 3 class includes the star clusters and galaxies expending in space, such as many Open clusters filled with young stars drifting apart. These young stars have been formed from the giant molecular clouds in which active star formation is occurring (Image 6). According to the Theory of Matrix, these star clusters and galaxies are 'young' in their energy scale, being in the process of their natural growth.

This classification applies to other multimodal objects and systems.

High energy massive radiating systems, such as stars and planets, might be referred to as the type 1 class of radiating objects and systems. They display Space-Time imbalance supported by Mass-Energy and Info-Energy imbalance. The Space-Time, Mass-Energy, and Info-Energy imbalance and associated influence of the Matrix forces indicate such properties of high energy massive radiating objects and systems as gravity prompted by the peripheral regions of these objects and systems and antigravity prompted by their centres.

The signs of antigravity, mainly associated with the past of the systems, such as the Earth or Omega Centauri, must be detectable at the systems' centres in the period of time now occurring. Space, masses, and energy, created under the influence of the periphery of the system, degenerate through the system under the impact of the system's centre. Please see my book 'Gravity and Antigravity to the Point' for details.

The specific rotation of the volume of the Earth or any other high energy massive radiating system about its Space-Time axis is prompted by the system's Space-Time, Mass-Energy, and Info-Energy imbalance and associated influence of the Matrix forces. It impacts on the rotation of the system about its axis of rotation in our dynamic world. The associated rotation of the Matrix grid, being simultaneously the external framework of the Earth in the multimodal Space-Time and its internal skeleton in our dynamic world, is reflected in the rotation of background radiation and must be detectable.

The type 2 class includes unbalanced non-radiating Space-Rising and toroidal systems. They also display Space-Time imbalance supported by Mass-Energy and Info-Energy imbalance. These systems expand and inflate spaces, masses, and energy and create one, two or several transmitting and non-transmitting Black Holes to obtain and retain the Space-Time, Mass-Energy, and Info-Energy balance.

The Space-Time, Mass-Energy, and Info-Energy imbalance and associated influence of the Matrix forces indicate such properties of the large spacious non-radiating objects and systems as antigravity prompted by the periphery of the systems and gravity prompted by their centres.

The specific rotation of a non-radiating system about its Space-Time axis is prompted by the system's Space-Time, Mass-Energy, and Info-Energy imbalance and associated influence of the Matrix forces. It impacts the system's rotation in our dynamic world. The associated rotation of background radiation must be detectable.

The type 3 class includes the star clusters and galaxies expending in space, such as many Open clusters filled with young stars drifting apart. These young stars have been formed from the giant molecular clouds in which active star formation is occurring.

Young stars, formed from the giant molecular clouds and drifting apart within Open clusters, may be referred to as the members of the type 3 class star clusters and galaxies. They expose the reverse of the maternal toroidal system. The non-radiating toroidal system, reaching the limits initiating a reverse (the Planck's units), is being reversed by the Matrix forces into the giant molecular cloud. The Planck's units, such as Planck length, Planck time, and Planck energy, reflect minimal conditions of an object or a system's existence and, accordingly, factors influencing the reverse of the objects and systems' Space-Time and Info-Energy flow.

The active star formation occurs as a result of the reverse of the direction of Space-Time and Info-Energy flow within the system. The reverse of the direction of the Space-Time and Info-Energy flow in a system will conduct the complete change of action in the system's multimodal Space-Time and Info-Energy structure - the system's Matrix.

The formation of volumes, masses and kinetic energy of the young stars reflects the balancing transformation of the Potential Space into the new-formed 3-dimensional structures, along with the transformation of the latent Info-Energy of the toroid into the actual Info-Energy of the young stars in the period of time now occurring on the late stages of the degeneration of a toroid in our dynamic world.

Image credit: NASA/JPL-Caltech/UCLA

BINARY AND MULTI-SYSTEMS

There is no empty space in the Universe. The 'outer space' is always another object or a system. Systemic understanding of nature provides new insights bringing together principles and concepts from different sciences and creating new patterns for the development of progressive models targeting explorations of the unknown.

Space, time, matter, energy, and information do not exist independent of objects and systems of the objects. They are the properties of the existing objects and systems. The qualities of the existing objects and systems are included, unified, and represented in the physical properties of the Universe as we know and sense it.

Every existing single object and every system of the objects, including visually 'empty' spaces, subatomic particles and holes, systems creating and holding radiating and gravitating Black Holes, other objects and systems existing in the Universe, are the multimodal objects and systems carrying the properties of different Space-Time modalities. Characteristics of an object or a system are different in different modalities of Space-Time.

Objects and systems, existing in the Universe, are tangled together by Space-Time and Info-Energy imbalance, building binary systems and multi-systems to obtain a balance. For example, an absorbing system, receiving volumes, masses, and energy from the outer space, is a part of a binary system or a multi-system. It is usually associated with a massive radiating system it feeds from.

The objects and systems, involved in a binary system or a multi-system's construction, might be distant in space. They are connected within their 2-dimensional Space-Time areas and central points of symmetry of each Matrix.

The system's development is determined by the structure of Space-Time and Info-Energy imbalance between the connected Matrixes and their associated systems.

Image credit: NASA/JPL-Caltech/STScI/Vassar

A binary system may be built by a high energy massive radiating system, such as a star or a radiating galaxy tangled by gravity, and a Space-Rising system, such as a toroid or another non-radiating system holding the Black Hole. These systems are connected within the 2-dimensional Space-Time region and central points of symmetry of each Matrix (Figure 3).

We would like to remind the reader that the exact reproduction of the Matrixes connection is impossible on the paper or in the computer simulation without damaging the main idea. Although we reproduce the correct form of the Matrixes, we rely entirely on the description.

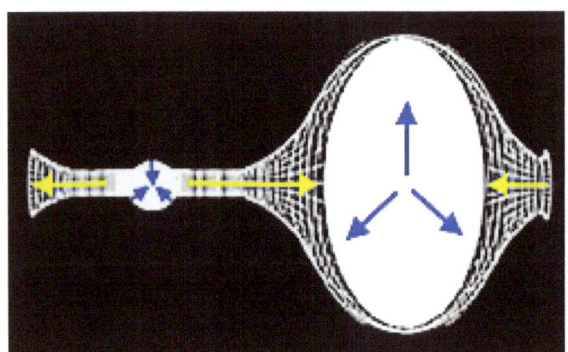

Figure 3: TRM associated with SRM

In this case, Arrows of Time indicate the direction of time and time-associated energy flow from the high energy massive radiating system, such as a star or a radiating galaxy tangled by gravity, into the toroid holding the central absorbing Black Hole.

Matter, volumes, energy, and related information may be drawn directly from the centre of the star - via the central points of the Matrix symmetry, the central Black Hole - into the centre of the associated toroid holding the Black Hole.

The volume and Info-Energy, bounded at the 3-dimensional centre of the TRM of the star, might be drawn indirectly, being adsorbed on to the filament of the 2-dimensional grid of the TRM and connected 'hungry' SRM of the toroid, and then transformed into the Potential Space and latent energy of the connected systems. The SRM's 2-dimensional grid performs, if necessary, the following absorption of the latent 2-dimensional Space-Time and Info-Energy and transformation into the inflated space and degenerated matter and energy of the non-radiating system that this SRM develops. Please see my book 'Black Holes' for details.

In case the massive radiating system has a large amount of total energy, the two associated Matrixes (TRM and SRM) and their systems may build a system with two centres, such as a Black Hole and a star, two bounded stars or galaxies.

As soon as the toroidal system reaches the minimal Space-Time and Info-Energy imbalance at the Matrix centre, it can be reversed.

Planck units, such as Planck length, Planck time, and Planck energy, reflect the minimal Space-Time and Info-Energy imbalance at the Matrix centres as the conditions that are necessary for the multimodal objects and systems to exist in our dynamic world.

Once the unbalanced Space-Rising non-radiating system, associated with the massive radiating system, such as a star or a radiating galaxy, is reversed, the direction of the Arrows of Time and time and energy flow is changed (Figure 4). The time and time-associated energy flow are directed from the Matrix centres to the points of the Matrixes connection.

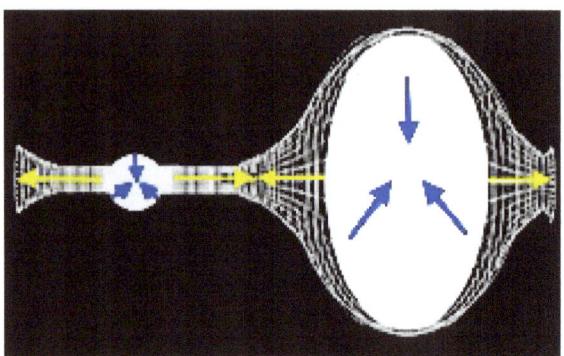

Figure 4: TRM associated with the reversed Matrix

If the massive radiating system keeps a large amount of total Info-Energy, the enormous flow of the excessive energy and associated information of the reversed system is blocked by the Info-Energy flow from the connected TRM.

Info-Energy of these connected systems may build a high degree compression upon the 2-dimensional grid in the areas of the Matrixes connection.

If the intermediate centre with '0' time-point characteristics is formed between the connected Matrixes, it is the weakest link in the system. Excessive energy may be released into 'outer space' (Figure 5).

The actual Info-Energy, creating a high degree compression, may damage the Info-Energy grid at the centre of the weaker Matrix. The excessive volumes, masses, and kinetic energy, bounded at the centre of the star, may create a high degree compression and destroy the 2-dimensional grid at the centre of the star shooting the jets of energy through the damaged areas of the grid.

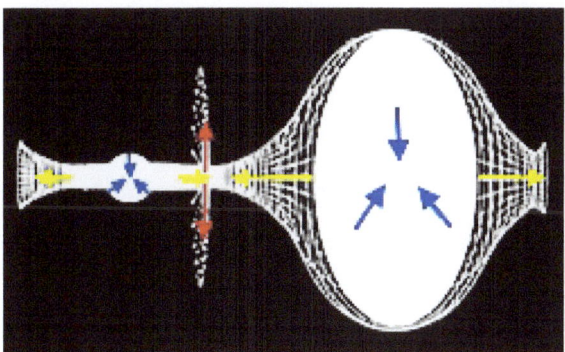

Figure 5: The strong TRM associated with the reversed Matrix

We can detect radiation, such as the cosmic microwave background radiation, and the release of the jets of the radiating energy before the enormous amount of volumes, matter, and energy breaks out via the central Black Hole.

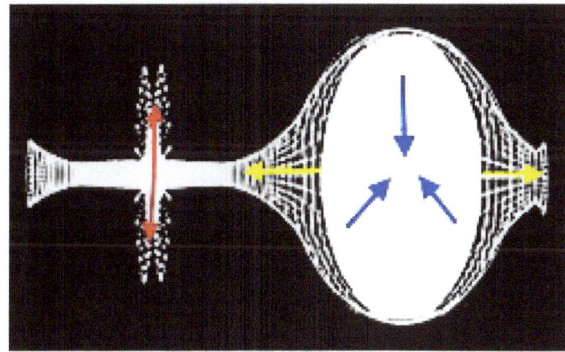

Figure 6: Supernova
The star explodes and releases the Supernova into our dynamic world (Figure 6).

Binary systems can be built by the connected SRMs of the large spacious and Space-Rising systems by contracting their attached Spaces of Time and central points of symmetry of each Matrix.

The large non-radiating systems, being in the process of integration, might be distant in space. They are connected within the 2-dimensional Space-Time region and '0' time-points of each Matrix. Accordingly, we can detect the Black Holes in the process of integration (Image 8).

In the early stage of integration, the deficit of the Potential Space and associated potential Info-Energy can damage the 2-dimensional Info-Energy grid in the area of the Matrixes connection if both SRMs and their associated systems are robust.

8

Image credit: NASA

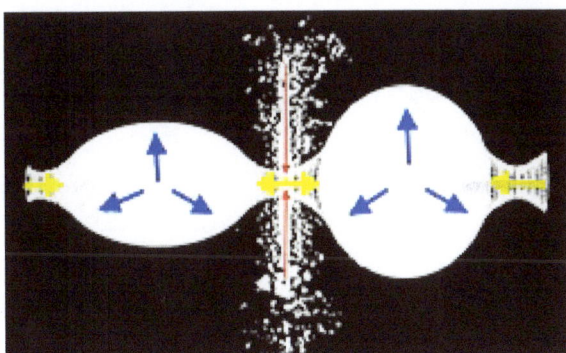

Figure 7: The early stage of integration of two Black Holes

Once the intermediate centre with the '0' time-point characteristics is formed in the damaged area, we can detect radiation, such as the cosmic microwave background radiation, indicating the following formation of a Black Hole in the 'visually empty' space or between Space-Rising galaxies. We can observe only matter that is drawn into this newly created Black Hole (Figure 7).

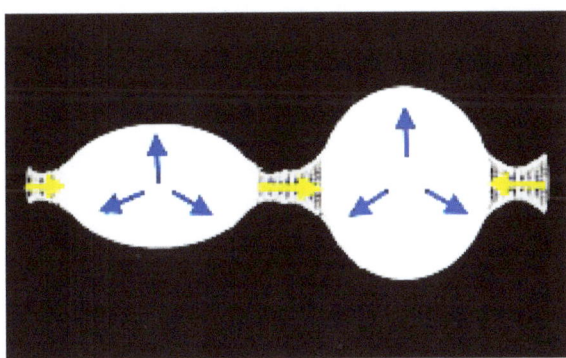

Figure 8: The later stage of the Black Holes integration

In the later stage of the Black Holes integration (Figure 8), the Space-Time and Info-Energy flow drains the weaker system, directly and indirectly, until the process of the systems integration is completed.

Image credit: DSS/NASA/JPL-Caltech

STIE-BLOCKS

There is no empty space in the Universe. The Universe is filled with the actual Info-Energy. We perceive this energy in different forms such as mass, kinetic energy, and other energy representations acting in the volume of the Universe in the period of time now occurring.

The 'outer space' is always the other object or the system. It can be a non-radiating system developing surface antigravity and space and mass degeneration and reduction prompted by the peripheral areas of the system. Alternatively, it is a tangled by gravity high energy massive radiating system generating consolidated spaces and masses prompted by the system's periphery.

No isolated objects and systems exist in our dynamic world. A balanced, isolated object or an isolated system in the state of equilibrium would cease to exist. Planck units, such as Planck length, Planck time, and Planck energy, reflect the minimal Space-Time and Info-Energy imbalance at the centres of the Matrixes as the conditions that are necessary for the multimodal objects and systems to exist in our dynamic world. The Arrows of Time are not harmonically balanced in the Matrixes of the existing objects and systems.

Multimodal radiating and non-radiating objects and systems, such as planets, stars, star clusters and galaxies, visually 'empty' spaces, subatomic particles and holes, quanta, systems holding Black Holes, other objects and systems existing in the Universe, are tangled together by the Space-Time, Info-Energy, and Mass-Energy imbalance reflected in gravity, antigravity, electric charges, and other characteristics.

Every Matrix displays a tendency to obtain and retain a balance and symmetry via the Space-Time axis and '0' Space-Time point at the centre of the Matrix. The excessive time and time-associated potential Info-Energy, along with the deficit of space and actual Info-Energy filling up the volume of the system with mass and kinetic energy, are intense in the central regions of the large unbalanced non-radiating systems, such as systems holding Black Holes and our Universe, and on the periphery of high energy radiating objects and systems, such as our planet and the Sun. It prompts consolidation and generation of volumes and masses - the process is reflected in gravitational acceleration.

The excessive space and space-associated actual Info-Energy, along with deficit of time and potential Info-Energy, are reflected in antigravity detectable on the periphery of the large spacious non-radiating systems, such as systems holding the central Black Holes, and at the centres of the high energy massive radiating systems, such as radiating galaxies, stars and planets, including the Sun and the Earth. It prompts degeneration and reduction of volumes and masses - the process is reflected in anti-gravitational deceleration.

The multimodal Matrixes provide a mechanism for the balancing Space-Time, Info-Energy, and Mass-Energy transformations.

The 2-dimensional grid, forming Matrix Spaces of Time, is a boundary of the enclosed object in the Matrix multimodal Space-Time. It creates the object's Potential Space in the form of the 2-dimensional 'container' surrounding the volume of the object in the Matrix Space of the Current Time.

The 2-dimensional grid of the Potential Space contains the object's potential energy, which is coded and fixed by the latent, or potential, information. The grid processes, stores, and transmits data along with energy. The description of its functioning of providing a regulatory layer between the time-associated latent Info-Energy and space-associated actual Info-Energy structures of the object would be similar to the function of structural genes and the membrane of a living cell.

The grid acts as a transmitter between the object's time-associated latent Info-Energy and its space-associated actual Info-Energy. The grid operates in a manner specific to the associated object/system, such as an electron or a star, in some similarity with the restricted genetic code expression, suppressing and altering the object's energy structures and keeping and transforming energy and associated information in compliance with a stored data. The 2-dimensional grid balances and enhances an existing capacity to perform work against its actual realisation in forms of actual Info-Energy, represented in mass, kinetic energy, and other forms of energy acting in the volumes of objects and systems in the period of time now occurring.

The object or the system's potential and actual Info-Energy structures counteract and keep a balance in the Matrix, retaining Space-Time, Info-Energy, and Mass-Energy Conservation Laws and Matrix Laws of Symmetry.

Balancing transformations are caused by the influence of the Matrix forces. The 2-dimensional grid vector-force of pressure and the object's vector-force of resistance act in a dynamic balance at the Matrix centre. The influence of these forces perform balancing Space-Time, Info-Energy, and Mass-Energy transformations. The Matrix 2-dimensional grid, forming the object's Potential Space, is the 'transformations horizon' in the process of the balancing transformations. Please see my books' Space and Time' and 'Energy of Existence' for details.

Radiation within the 2-dimensional Space-Time is not the emission or progression but a divergence activating Matrix vector-forces. The quantity of flux does not emanate from the point of symmetry but diverge from the point of a balance - the '0' Space-Time-Energy point of the object's multimodal Space-Time and Info-Energy structure. Similarly, +x requires -x to establish zero. The balancing Space-Time and Info-Energy' flow' is the divergence between the unbalanced elements of the object's different modalities. Accordingly, the Space-Time and Info-Energy 'flow' within the object's 2-dimensional structure affects the Space-Time and Info-Energy flow within the volume of the object in our dynamic world.

It is reflected in changes in volume, mass, kinetic energy, shape, and structure of the object. It affects magnetic fields, electric currents, the intensity of gravity and antigravity, and brings about the object's qualitative and quantitative changes.

Space-Time, Info-Energy, and Mass-Energy transformations propagate with the speed of light. 'Relativity theory ... shares with the corpuscular theory of light the unusual property that light carries inertial mass from the emitting to the absorbing object.' [Einstein Albert, The Development of Our Views on the Composition and Essence of Radiation (1909)].

Background radiation and light carry inertial mass from the emitting to the absorbing object within the system and from the emitting to the absorbing system within the larger system, integrating the existing objects and systems into the multimodal Universe. Background radiation carries Space-Time, Info-Energy, and Mass-Energy transformations with the speed of light.

Although, the 2-dimensional representations of background radiation, arranging the 2-dimensional grids of the Matrixes of the existing objects and systems, interrelate with dynamic forms of background radiation of our dynamic world. Dynamic representations of background radiation build the dynamic internal framework and Space-Time and Info-Energy skeletons of the existing objects and systems. This interrelationship provides a mechanism for the Space-Time, Info-Energy, and Mass-Energy transformations and their transmission in our dynamic world.

Dynamic representations of background radiation arrange, support, transport, and transmit the Space-Time, Info-Energy, and Mass-Energy transformations, including those associated with gravity and antigravity.

The speed of gravity propagation was introduced by Hendrik Lorentz (1900) and confirmed by Hermann Minkowski, 'The law of mass attraction, which has been just described and which is formulated in accordance with the relativity postulate, would signify that gravitation is propagated with the velocity of light.' [Minkowski Hermann, The Fundamental Equations for Electromagnetic Processes in Moving Bodies, Appendix (1908)]

In the Theory of Matrix, we develop the theory of light, uncover unusual property of light, and introduce the Coefficient of Transformation $[c]^2$, applicable to the decisions associated with Space-Time, Info-Energy, and Mass-Energy transformations, including those associated with gravity and antigravity. The Coefficient of Transformation $[c]^2$ equals the numerical value of the speed of light squared.

The frequency of the electromagnetic waves and their wavelength influence the object's 2-dimensional Info-Energy grid, changing Space-Time, Mass-Energy, and info-Energy properties of the objects and systems via the energy transfer. Dynamics preserve the Theory of Matrix natural laws, such as the Laws of Space-Time, Mass-Energy, and Info-Energy Transformation, Reversibility, Conservation, Limitation, and Symmetry, which are described in the Theory of Matrix series of books. These principles are responsible for the harmony of the existing Universe.

The speed of light defines the dynamic body of the multimodal Universe. The complex Space-Time and Info-Energy structure of background radiation make multimodal Matrixes the STIE-blocks (Space-Time-Info-Energy building blocks) of the Universe. Objects and systems, existing in the Universe, are tangled together by Space-Time and Info-Energy imbalance and balancing transformations. Dynamic representations of background radiation carry the latent, or potential, Info-Energy structure of our dynamic world.

Anisotropies, or irregularities, of background radiation, such as the irregularities of the cosmic microwave background radiation and infrared background, reflect the reorganisation of space, time, matter, energy, and information in the Universe. The process reflects the two opposite forces operating at the centres of the multimodal Matrixes of the Universe and other existing objects and systems. They perform the balancing Space-Time, Info-Energy, and Mass-Energy transformations related to the process of generation and degeneration of the multimodal Space-Time and Info-Energy structures existing in our dynamic world.

THE MULTIMODAL UNIVERSE

Time, space, mass, energy, associated information, forces, other characteristics and transformations of the multiple Universe modalities are integral elements of the greater multimodal Universe that we call Meta-Universe. The Meta-Universe as a system utilises the properties of our thermodynamic modality of the Universe

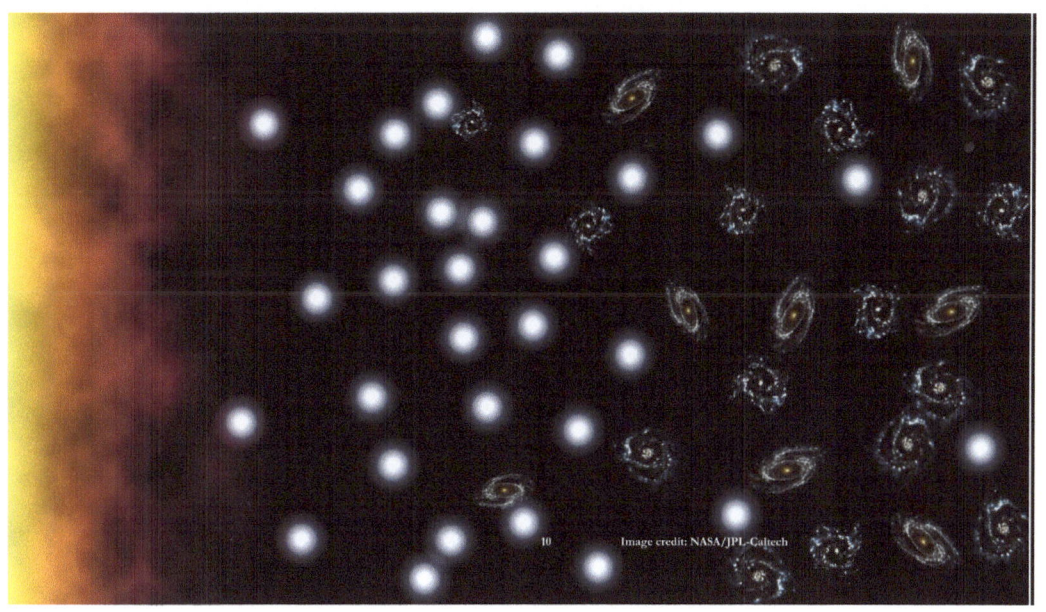

and other co-existing modalities.

The interrelation of the Universe modalities

Our thermodynamic Universe is young, growing, temporarily 'inflating', non-radiating Universe. The Space-Time imbalance and associated Info-Energy imbalance display space, mass, and energy degeneration and reduction reflected in antigravity in the peripheral regions of the Universe. Simultaneously, space and mass generation is reflected in gravity at the centre of the Universe.

Our dynamic Universe modality is associated with the robust radiating modality of the Meta-Universe. They are connected within the 2-dimensional Space-Time region and '0' time-points of each Matrix (Figure 9). Accordingly, the two associated systems build a system with two centres.

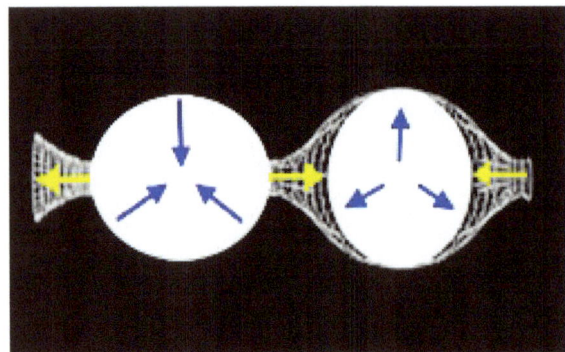

Figure 9: The Primary Giant Black Hole of the Universe

The Primary Giant Black Hole is being built by the radiating modality of the

Meta-Universe and the absorbing modality of our dynamic world. This Black Hole within the 2-dimensional Space-Time region is similar to the two sides of the same 2-dimensional coin.

The Primary Giant Black Hole is a contact path equilibrating the system. Our Universe 'breathes in' and receives the Space-Time and Info-Energy through this giant central Black Hole. The development of this binary system is determined by the Space-Time and Info-Energy imbalance of the connected systems.

Volumes, energy, and masses, generated under the influence of the Space-Time and Info-Energy imbalance, such as excessive time and time-associated latent energy and deficit of space and mass at the centre of the Universe, degenerate through the system into the Dark matter and Dark energy under the influence of the Space-Time and Info-Energy imbalance on the periphery of the Universe.

The radiating modality of the Meta-Universe affects the rotation of our non-radiating modality about its Space-Time axis that influences its associated rotation about the axis of rotation and precession in our dynamic world.

The relative rotation of the Matrix grid has a strong effect on the rotation of background radiation, such as the cosmic microwave background radiation considered in relation to the body of the Universe, and its gradual shift in the orientation of the axis of rotation that must be detectable.

The process of our thermodynamic Universe growth may be reversed. If our Space-Rising Universe develops the critical form of the 2-dimensional toroid and reaches the limits established by Max Planck (Planck length, Planck time, and Planck energy), it will be reversed under the influence of the Matrix forces changing properties of our Universe and the direction of the time and energy flow within the Black Hole.

If the toroidal Universe is reversed, the Potential Space and associated latent 2-dimensional Info-Energy of the Matrix grid build high degree pressure on the Black Hole structure. This continuous pressure is the measure of the excessive latent, or potential, Info-Energy of the toroidal Universe. The Matrix grid vector-force of pressure develops the flow of the Potential Space, created volumes, energy, and matter via the Black Hole to the central areas of the complement to our dynamic world - the associated modality of the Meta-Universe, and the Black Hole 'radiates' within the Meta-Universe.

Our Universe will 'breathe out' the excessive Space-Time and Info-Energy through this giant central Black Hole, creating a Supernova in the associated Universe. Please see my book 'Black Holes and Supernovas' for details related to the Supernova release.

Following our current perception of the world, we have to consider the multimodal Meta-Universe as a compound system and, accordingly, the Meta-Universe modalities as compounds.

Our Universe might participate in the creation of a binary system of Black Holes built by the connected non-radiating modalities of the Meta-Universe by contracting their connected Spaces of Time and central points of symmetry of each Matrix.

The binary system of Black Holes would reflect the process of integration of the modalities of the Meta-Universe. We, humans, would not notice the integration of our Universe with another supergiant modality of the Meta-Universe but the changed properties of our world.

Every modality of the Meta-Universe has the start and the end in time and space. They are limited in space, time, mass, energy, and information.

Different modalities of the Universe, including Multiverse, our Universe and other Universes, stages of the Universe development, and other modalities of the Great Universe, co-exist in the state of Equilibrium in the Multimodal World. They are unified, balanced, and reflected in the resultant force of the Meta-Universe.

The resultant force of the Meta-Universe equals zero.

The Universe, combined into one, whole, includes all existing matter, energy, time and space considered as a well-ordered whole. It is the cosmos - the Multimodal world. We call it 'Meta-Universe' denoting the Universe as a multimodal system combined into one.

The Meta-Universe exists in the state of Equilibrium. The properties of the multiple modalities are nullified in the balanced Meta-Universe. Accordingly, the Great Universe was not 'born'. It has no start of the existence in space, time, mass, energy, and information. It does not have the end but modalities.

Image credit: NASA/JPL-Caltech

THE MULTIMODAL WORLD

FOUR PRINCIPLES OF THE OBJECTS EXISTENCE

1. The existence of objects and systems rises in the Space-Time, Info-Energy, and Mass-Energy imbalance. A balanced, isolated object or an isolated system in a state of equilibrium would cease to exist.

Planck length, Planck time, and Planck energy are to be understood as the minimal Space-Time and Info-Energy imbalance and, accordingly, the minimal conditions of an object or a system's existence in our dynamic world.

The existence of the objects and systems is preserved by the Natural Laws, such as the Laws of Space-Time, Info-Energy, and Mass-Energy Conservation, Reversibility, Transformations, Limitation and the Laws of Symmetry, described in the Theory Of Matrix.

2. Objects and systems, existing in the Universe, are tangled together by the Space-Time and Info-Energy imbalance and balancing transformations.

3. Matrixes represent the multimodal Space-Time and Info-Energy structures of the existing objects and systems. The Matrix grid is being simultaneously the external framework of an object or a system in the multimodal Space-Time and its internal skeleton in our dynamic world.

4. Background radiation supports the Info-Energy grid of the Matrixes for the Universe and the existing objects and systems. Any object, which is smaller than the wavelengths, supporting the 2-dimensional grid of its Matrix, does not exist in Space-Time. The complex Space-Time and Info-Energy structure of background radiation make multimodal Matrixes the Space-Time-Info-Energy building blocks (STIE-blocks) of the Universe.

Image credit: NASA/JPL-Caltech/UCLA

PROSPECTS

The Theory of Matrix applies to every object and every system of the objects, including visually 'empty' spaces, subatomic particles and holes, quanta, humans, Black Holes, other existing objects and systems. The Theory of Matrix applies to the Universe and its Matrix.

The application of the Theory of Matrix provides an opportunity to access some additional Space-Time characteristics and properties of the multimodal objects and different regions of our Universe, their Space-Time, Mass-Energy, and Info-Energy structures; Space-Time, Mass-Energy, and Info-Energy transformations, characteristics of gravity, antigravity, and specific rotation of cosmic background radiation.

Energies of background radiation, including those known as cosmic background radiation, such as the cosmic microwave background radiation, and light and associated information, supporting the Info-Energy grid of the multimodal Matrixes, make Matrixes the STIE-blocks of the Universe.

The dynamic representations of background radiation are accessible for our investigation. The study of background radiation and its energy characteristics in the 2-dimensional Space-Time settings will allow to gain additional knowledge associated with the structure of the Universe and objects existing in the Universe and provide a vast spectrum of opportunities for a broad range of the natural sciences.

'I'm of the opinion that physics is conceptual, not illustrative.' [(Albert Einstein, The Bad Nauheim Debate (1920)]

Conceptual physics (plural noun, treated as sing.) is the study and description of properties and interactions of space, time, matter, and energy in a conceptual non-mathematical form with emphasis on logical reasoning in order to derive fundamental laws of nature and obtain conclusions from these laws by a sequence of logical steps. It is the branch of theoretical physics.

The human perception of time is reflected in mathematics and other abstract sciences associated with numbers, quantities, and space. It affects the theoretical concepts, which are built on mathematics and applied to other disciplines, such as physics. Mathematics based on single numbers, where 1^3 still equals 1, will not help to overcome the barrier of pointed consciousness. Development of 3-dimensional mathematics is necessary.

The Theory of Matrix supports the next step of the Space-Time theory development concerning the 3-dimensional time, its practical application to the 3-dimensional space and the following merger into the new quality of 3-dimensional Space-Time understanding.

ABOUT THE AUTHOR

The Theory of Matrix is a new theory combining the elements of psychology, cosmology, and astrophysics. This theory was introduced by Dr Audrey Elizabeth Randles in her work 'The Theory of Matrix' in 2012.

Dr Randles developed the program for psychological investigations of the Universal Matrix along with the development of the Coresynthesis Psychological Model in early 1990s. There was a space of ten years between the first explanations of the results and association of these results with the characteristics mathematically introduced by Lorenz, Minkowski, and Einstein for the Light Cone.

When Dr Randles published her work, the Light Cone was considered a specific case applicable to the flash of light. She brought forward the idea of the universality of the Space-Time structure of the objects and a new vision on Space-Time physics. Following the analysis of the parallels and variations between the Universal Matrix and the Light Cone, Dr Randles calls the Light Cone 'the Matrix of Light' and presents the Theory of Matrix as the relative importance for a true understanding of the multimodal world.

The Theory of Matrix introduces a new understanding of the multimodal world with respect to multidimensional time. Time is no longer seen as a dimension of space, nor as a momentary feature of an event, but as a multidimensional element in its own right.

Dr Randles associates space with the actuality and time with the potentiality or latency. Therefore, the objects are viewed as multimodal, multidimensional objects in Space-Time.

Books on Cosmology by Audrey E. Randles:

'Black Holes and Supernovas' (2016)

'Meta-Universe' (2016)

'Antigravity' (2015)

'Supernovas' (2015)

'The Giant Black Hole of the Universe' (2015)

'Energy in Cosmology' (2014)

'Gravity and Antigravity to the Point' (2014)

The Theory of Matrix series of books (2012 - 2013):

'Blocks of the Universe'

'Space and Time'

'Energy of Existence'

'Gravity and Rotation'

'Black Holes'

'Matrix of the Universe'

The New 2020 books on Cosmology include Kindle eBook and a paperback of the same title at Amazon's Book Store.

THE MULTIMODAL WORLD

THE LIST OF THE IMAGES

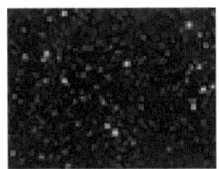

Image 1: 'Take a Splash Into the Cosmos'
Image date: SEPTEMBER 9, 2014
Image credit: NASA/JPL-Caltech

'Millions of galaxies populate the patch of sky known as the COSMOS field, short for Cosmic Evolution Survey, a portion of which is shown here. Even the smallest dots in this image are galaxies, some up to 12 billion light-years away. The square region in the center of bright objects is where the telescope was blinded by bright light. However, even these brightest objects in the field are more than ten thousand times fainter than what you can see with the naked eye.' NASA

Image 2: 'Ultraviolet Extensions'
Image date: APRIL 16, 2008
Image credit: NASA/JPL-Caltech/MPIA

'This ultraviolet image from NASA's Galaxy Evolution Explorer shows the Southern Pinwheel galaxy, also know as Messier 83 or M83. It is located 15 million light-years away in the southern constellation Hydra.' NASA

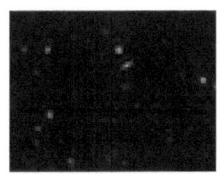

Image 3: 'NEOWISE Opens its Eyes'
Image date: DECEMBER 19, 2013
Image credit: NASA/JPL-Caltech

'It shows a patch of sky in the constellation Canes Venatici, or the Hunting Dogs. The galaxy seen near the center is NGC 4111, the largest member of a small group of galaxies located 50 million light-years away. The galaxy is similar in brightness to our Milky Way galaxy, but with a diameter of 70,000 light-years, it is slightly smaller. NGC 4111's core contains a supermassive black hole actively feeding off surrounding gas and dust.' NASA

Image 4: 'Masking Out Galaxies'
Image date: NOVEMBER 6, 2014
Image credit: NASA/JPL-Caltech

'This graphic illustrates how the Cosmic Infrared Background Experiment, or CIBER, team measures a diffuse glow of infrared light filling the spaces between galaxies. The glow does not come from any known stars and galaxies; instead, the CIBER data suggest it comes from stars flung out of galaxies.' NASA

Image 5: 'NASA's Spitzer 12th Anniversary Space Calendar'
Image date: AUGUST 20, 2015
Image Credit: NASA/JPL-Caltech

'NASA's Spitzer Space Telescope celebrated its 12th anniversary with a new digital calendar showcasing some of the mission's most notable discoveries…' NASA

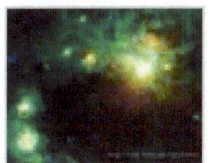

Image 6: 'Star Formation Everywhere You Look'
Image date: JUNE 24, 2011
Image credit: NASA/JPL-Caltech/UCLA

'There are five distinct centres of star birth in this one image alone. Star-forming nebulae (called HII regions by astronomers) are clouds of gas and dust that have been heated up by nearby stars recently formed from the same cloud... Going counter-clockwise from Gum 22, the other catalogued nebulae in the image are Gum 23 (part of same cloud as 22), IRAS 09002-4732 (orange cloud near center), Bran 226 (upper cloud of the two at lower left), and finally Gum 25 at far lower left.' NASA

Image 7: 'Eyes in the Sky'
Image date: APRIL 26, 2006
Image credit: NASA/JPL-Caltech/STScI/Vassar

'These shape-shifting galaxies have taken on the form of a giant mask. The icy blue eyes are actually the cores of two merging galaxies, called NGC 2207 and IC 2163, and the mask is their spiral arms... NGC 2207 and IC 2163 met and began a sort of gravitational tango about 40 million years ago.' NASA

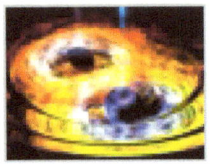

Image 8: 'Two Black Holes on Way to Becoming One (Artist's Concept)'
Image date: DECEMBER 3, 2013
Image credit: NASA

'Supermassive Black Holes at the hearts of galaxies are thought to form through the merging of smaller, yet still massive Black Holes, such as the ones depicted here.' NASA

Image 9: 'A New Solar Neighbour'
Image date: MARCH 7, 2014
Image credit: DSS/NASA/JPL-Caltech

'A nearby star stands out in red in this image from the Second Generation Digitised Sky Survey... The star WISEA J204027.30+695924.1 is a dim star called an L-subdwarf, and is particularly fast moving most likely because it's old. Older stars tend to have more time -- billions of years -- to get flung around, and pick up speed.' NASA

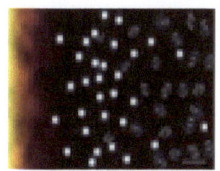

Image 10: 'Baby Galaxies in the Adult Universe'
Image date: DECEMBER 21, 2004
Image credit: NASA/JPL-Caltech

'This artist's conception illustrates the decline in our universe's 'birth-rate' over time. When the universe was young, massive galaxies were forming regularly, like baby bees in a bustling hive. In time, the universe bore fewer and fewer 'offspring,' and newborn galaxies (white circles) matured into older ones more like our own Milky Way (spirals).' NASA

Image 11: 'North America Nebula in Different Lights'
Image date: FEBRUARY 10, 2011
Image credit: NASA/JPL-Caltech

'This new view of the North America nebula combines both visible and infrared light observations, taken by the Digitized Sky Survey and NASA's Spitzer Space Telescope, respectively, into a single vivid picture. The nebula is named after its resemblance to the North American continent in visible light, which in this image is represented in blue hues. Infrared light, displayed here in red and green, can penetrate deep into the dust, revealing multitudes of hidden stars and dusty clouds.' NASA

Image 12: 'A Black Hole in Omega Centauri'
Image date: AUGUST 16, 2010
Image credit: NASA/JPL-Caltech/UCLA

'The ancient astronomer Ptolemy thought Omega Centauri was a star, and Edmond Halley identified it as a nebula in 1677. In the 1830s, John Herschel identified it as a globular star cluster orbiting our Milky Way galaxy. A globular cluster is a spherical group of stars that are bound together by gravity.' NASA

THE MULTIMODAL WORLD

Content Use Policy

© Audrey Elizabeth Randles

Content may be used for any purpose without prior permission, subject to the special cases noted below.
By downloading the material the user agrees:
1. to use a credit line in connection with the content. Unless otherwise noted in the caption information for any content and images the credit line should be as follows:
'Audrey E. Randles, The Multimodal World (2020)'
2. that we do not represent others who may claim to be authors or owners of copyright of any of the content, and make no warranties as to the quality of the content;
3. that we shall not be responsible for any loss or expenses resulting from the use of the content, and you release and hold us harmless from all liability arising from such use.
Special Cases:
This content is available for educational, journalistic, personal uses, and scientific research in accordance with a scientific code of ethics. Restrictions are placed on commercial uses. To obtain permission for commercial use, contact the copyright owner by email coresynthesis@gmail.com

www.ingramcontent.com/pod-product-compliance
Lightning Source LLC
Chambersburg PA
CBHW051207220526
45473CB00003B/941